YOUR KNOWLEDGE HAS VALUE

- We will publish your bachelor's and master's thesis, essays and papers

- Your own eBook and book - sold worldwide in all relevant shops

- Earn money with each sale

Upload your text at www.GRIN.com and publish for free

Bibliographic information published by the German National Library:

The German National Library lists this publication in the National Bibliography;
detailed bibliographic data are available on the Internet at http://dnb.dnb.de .

This book is copyright material and must not be copied, reproduced, transferred, distributed, leased, licensed or publicly performed or used in any way except as specifically permitted in writing by the publishers, as allowed under the terms and conditions under which it was purchased or as strictly permitted by applicable copyright law. Any unauthorized distribution or use of this text may be a direct infringement of the author s and publisher s rights and those responsible may be liable in law accordingly.

Imprint:

Copyright © 2018 GRIN Verlag
Print and binding: Books on Demand GmbH, Norderstedt Germany
ISBN: 9783668837652

This book at GRIN:

https://www.grin.com/document/447110

Masood Kaloo, Bilal Bhat

An Introduction to Molecular Recognition Approach. Tool for Environmental Analysis

GRIN Verlag

GRIN - Your knowledge has value

Since its foundation in 1998, GRIN has specialized in publishing academic texts by students, college teachers and other academics as e-book and printed book. The website www.grin.com is an ideal platform for presenting term papers, final papers, scientific essays, dissertations and specialist books.

Visit us on the internet:

http://www.grin.com/

http://www.facebook.com/grincom

http://www.twitter.com/grin_com

An Introduction to Molecular Recognition Approach: Tool for Environmental Analysis

Authors

MASOOD AYOUB KALOO
BILAL AHMAD BHAT

Department of Chemistry, Govt. Degree College, Shopian, J&K-192303 (INDIA)

Introduction

In the past few decades, an increase in the number of various chemical species (gases, anions, cations, organic and inorganic compounds) has resulted in the degradation of the natural environment.[1-3] Many chemical species in different environmental sub-systems are responsible for various detrimental effects, including degradation of water and air.[4-7] These detrimental effects continue to accelerate due to urbanization and fossil fuel derived energy generation.[8-10] Thus, there is a growing need to monitor the environment and the sources of contaminants to the environment, in order to control pollution and prevent its rise in the future.[11-13]

Development of new and convenient approaches to monitor environmental pollution and understanding of environmental processes is imperative.[14-16] In this direction, advances in method development for environmentally significant analytes is vital.[17-19] Molecular recognition presents one such advancement via analytical tools in the form of molecular receptors.[20-26] Such approaches provide rapid and naked-eye based signaling response. Most importantly, they are easy to use and provide a cost-effective means of analysis.[27-29]

Molecular recognition and receptor approach

Molecular recognition events yield information about environmental analytes at molecular level through specific binding of substrate by a molecular receptor. Such binding is regulated by geometrical and electronic complementarity between receptor and analyte.[30-33] The receptor can be defined as a molecular entity of abiotic origin that interacts with the analyte (anion, cation or a neutral molecule) and offers a unique response (Figure 1), with a simultaneous signal transduction in the form of photophysical or redox signaling.

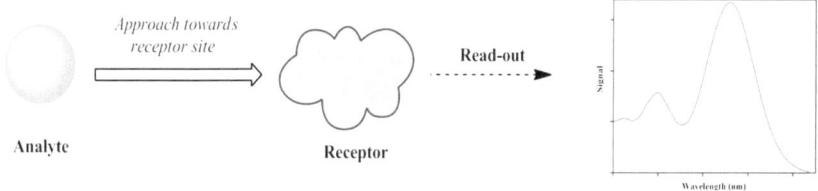

Figure 1 Presentation of molecular receptor approach.

Usually, molecular recognition events employ two different strategies to identify molecules, depending upon the mode of binding between the receptor (usually a big molecule) and the analyte (usually a minute structure).[34] These two strategies have been categorized as chemosensing and chemodosimetry. On one hand, chemosensors utilize interaction of target analyte with the receptor through a non-covalent interaction in order to yield a measurable optical signal with real-time response, usually within seconds. Such basis of recognition offers them with the advantages of reusability or recyclability. On the other hand, chemodosimetry involves making and breaking of bonds, and events are generally slow and irreversible.[35]

Chemosensensing

Chemosensing utilizes reversible interactions *via* a binding-site signaling sub-unit or through displacement approach.[36] In binding-site signaling approach, a linker (usually aliphatic chain) separates signaling unit (chromophore or a fluorophore) from the target site of the analyte.[36-39] This can further be tailored by making receptor site itself a part of signaling unit, so as to generate effective read-out behavior and hence achieve high sensitivity (Figure 2). The approach operates through hydrogen-bond donation from receptor to analyte or proton transfer signaling. In addition to this, such receptors also involve halogen bonding, *pi*-interactions and hydrophobic interactions.[40]

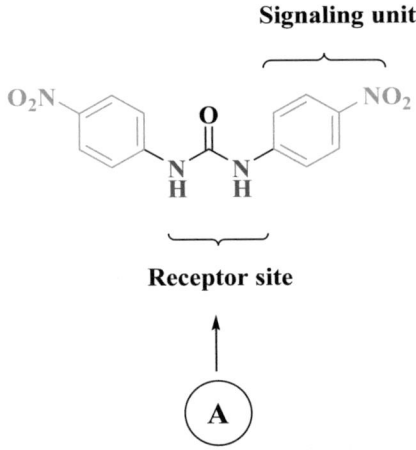

Figure 2 Analyte (**A**), approaching towards the receptor site, in binding site-signalling approach.

In the binding site-signaling approach, Liu *et al.* reported ratiometric fluoride anion recognition via a proton transfer signaling (PTS) with 4-benzoylamido-N-butyl-1, 8-napthalimide receptor **1.1** (Figure 3).[41] Interaction of fluoride at the receptor site concentrated negative charge on the amide nitrogen and caused a red-shift in the absorption spectra.

Similarly, measurable changes were observed in the emission properties of the fluorophore. These events were observed in the form of striking colourless to yellow and blue to orange emission signals. Similarly, Mukhopadhyay *et al.* produced panchromatic recognition for CN⁻ anion and (Cu^{2+}/Fe^{2+}) cations by electron transfer between analyte and electron deficient naphthalene diimides **1.2** (Figure 3).[42] The existence of persistent anion as well as cation radicals were evaluated through absorption (UV-Vis-NIR) and emission spectroscopy. The corresponding recognition events were noticed via both chromogenic display as well as fluorogenic quenching in the solution.

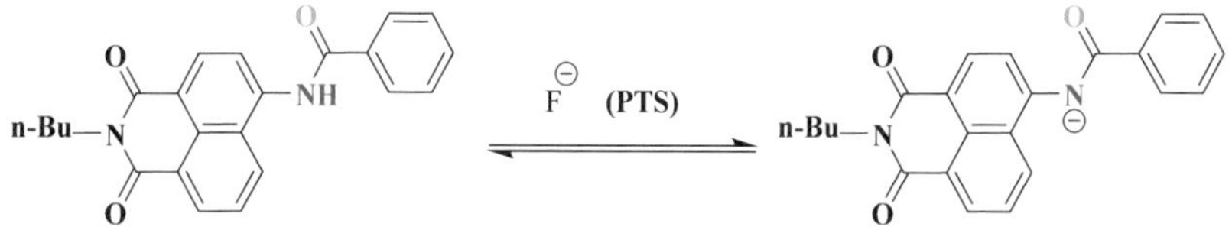

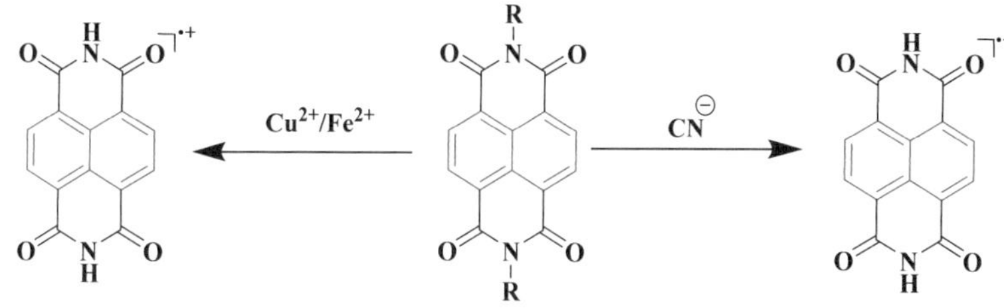

Figure 3 Molecular recognition through binding site-signalling approach. **1.1** receptor demonstrates proton transfer signalling (PTS) for fluoride, while as **1. 2** involves electron transfer phenomenon.

In the displacement approach both the reporter as well as binding site are free from any covalent linkage, and they become an ensemble through formation of a coordination complex (Figure 4).[43] The approach usually involves competition between two different types of guests for the ensemble formation. Here, initially existing complex upon impact of target species, dislodges and forms a new complex with the incoming guest. The corresponding displacement reaction is marked by signal transduction. The most common interactions between receptor and analyte involve hydrogen bonding, electrostatic forces, and coordination bond with metal centers or electron transfer phenomenon. The interaction is further controlled by the geometry of the guest, charge, hydrophobicity and solvent system under operation. In displacement approach, Fabbrizzi et al. reported an ensemble of copper (II) and coumarin fluorophore (**F**) **1.3** (Figure 4), for HCO_3^- recognition. The complexation initially shows fluorescence quenching. However presence of analyte with Y-shaped bite (carbonate, acetate) replaces the fluorophore and in turn promotes fluorescence enhancement for detection of analyte in aqueous media.[44]

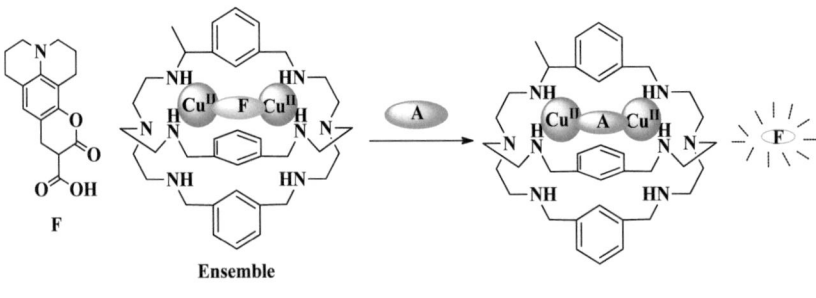

Figure 4 Molecular recognition through displacement approach. Here **A** refers to target analyte and **F** the initially bound guest.

Chemodosimetry

The approach utilizing a covalent linkage between receptor and analyte usually produces a specific and irreversible chemical structure. This approach was named as chemodosimetry, for the first time by Chae and Czarnik. They described a chemodosimeter as abiotic molecule used for analyte recognition with the concomitant irreversible transduction during recognition events.[45] The interaction between target analyte (anion, cation, neutral molecule) and the molecular probe involves significant chemical transformation via breaking and forming of covalent bonds (Figure 5). The outcome of all these events results in the formation of products with different photophysical properties and thus, permits the chemical characterization of the analyte under investigation.[46] These events are usually irreversible, but can be reversed in a few cases through a specific transformation, which is different from that involved in target analyte recognition. These chemodosimetric events are less affected by the environment compared to chemosensing approaches. Unlike chemosensors, chemodosimetry involves analyte recognition, either by direct chemical reaction of analyte with a chemodosimeter to form a new product, or by catalysis of a chemical reaction, which results in the modification of the chemodosimeter with a concomitant change in its optical characteristics. For example, in the first case, Kim et al. reported selective fluorescent chemodosimeter **1.4** (Figure 5), for cyanide anion recognition.[47] Here coumarin acted as fluorescent signaling unit and salicylaldehyde as a receptor site. The nucleophilic addition of CN^- phenolic on the aldehyde functionality promoted the formation of tetrahedral carbon centre and thus disrupted the initially existing hydrogen bond between carbonyl and phenolic hydrogen. After nucleophilic attack, **1.4** underwent a fast proton transfer from phenol hydrogen to alkoxide. These events ended up with fluorescence "turn on" as output signal to detect cyanide anion. In the second case (i.e. catalysis of a reaction), Chang et al. developed a thiocoumarin **1.5** (Figure 5), and mercury cation (Hg^{2+}) based chemodosimetry for naked-eye detection of Hg^{2+} in acetonitrile-water system.[48] Here presence of Hg^{2+} ion induces transformation of thiocoumarin to coumarin. During this event colorimetric changes from pink to yellowish green were achieved along with fluorescence enhancement.

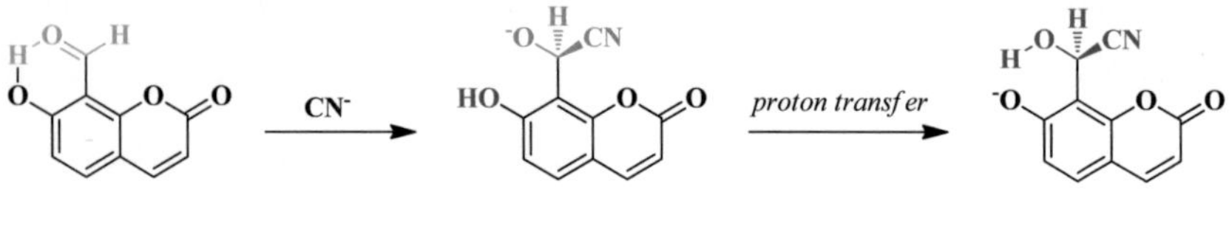

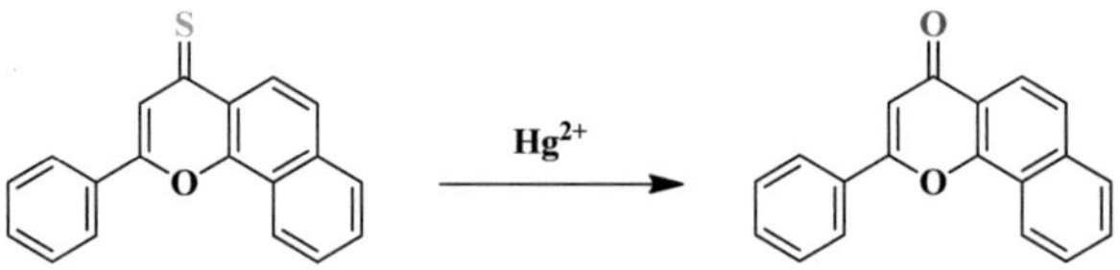

Figure 5 Various chemodosimetric approaches for analyte recognition.

Conclusions

From the above discussion, it is clear that the field of molecular recognition which is popular since 1987 (noble prize in supramolecular chemistry), holds enough promise and scope for the development of miniature, cheap and sensitive molecular receptors for environmental analysis. Despite some concerns regarding the understanding of the nature of interactions (non-covalent in particular) during the recognition events, read-out performances of such methods make them attractive tools to monitor environmental analytes. As far as involvement of receptor technology in characterizing diverse chemical species are concerned, it is reasonable to point out that research is still at birth stage. Even though various researchers have pointed out their design molecular systems with futuristic applications, however the system lack real time applications and are workable under lab conditions only. Thus, researches in this area when it comes towards the development of molecules deciphering real-time promise are highly demanding and crucial.

Acknowledgements

We thank the honourable principal and department of chemistry (GDC Shopian) for their support for writing this book chapter. M. A. Kaloo gratefully acknowledges Department of Science and Technology, New Delhi for INSPIRE-FACULTY award [DST/INSPIRE/04/2016/000098].

References

1. Vitousek, P. M.; Mooeny, H. A.; Lubchencon, J.; Melillo, J. M. *Science* **1997**, *277*, 494.

2. Syvitski, J. P. M.; Vorosmarty, C. J.; Kettner, A. J.; Green, P. *Science* **2005**, *308*, 376.

3. Crutzen, P. J. *Faraday. Discuss.* **1995**, *100*, 1-21.

4. Aguinaga, N.; Campillo, P.; Vinas, P.; Hernandez-Cordoba. *Anal. Chem. Acta.* **2007**, *596*, 285.

5. Goldewijk, K. K. *Biogeochem. Cycles* **2001**, *15*, 417.

6. De Bruyn, A. M. H.; Gobas, F. A. P. C. *Ecol. Model.* **2004**, *179*, 405.

7. Kirchner, J. W.; Neal, C. *Anal. Chem. Acta.* **2013**, *110*, 12213.

8. Ghermandi, A.; Bixio, D.; Thoeye, C. *Sci. Total. Environ.* **2007**, *380*, 247.

9. Oster, J. D.; Whichels, D. *Irri. Sci.* **2003**, *22*, 107.

10. Costello, C.; Griffen, W. M.; Matthews, H. S.; Weber, C. L. *Environ. Sci. Technol.* **2011**, *45*, 4937.

11. Jones, N. *Nature* **2009**, *458*, 1094.

12. Dahmus, J.; Gutowski, T. *Environ. Sci. Technol.* **2007**, *41*, 7543.

13. House, K.; Harvey, C.; Aziz, M.; Schrag, D. *Energy. Environ. Sci.* **2009**, *2*, 193.

14. El-Shaarawi, A. H. *Mathmetics and Computers in Simulation.* **1995**, *39*, 441.

15. El-Shaarawi, A. H.; Niculescu, S. P. *Environmetrics* **1992**, *3*, 389.

16. El-Shaarawi, A. H.; Niculescu, S. P. *Environmetrics* **1993**, *4*, 233.

17. Lin, P.; Yan, F. *Adv. Matter.* **2012**, *24*, 34.

18. Majid, S.; Rhazi, M. E.; Amine, A.; Curulli, A.; Palleschi, G. *Microchim. Acta.* **2003**, *143*, 195.

19. Somerset, V.L.; Leaner, J.; Mason, R.; Iwuoha, E.; Morrin, A. *Int. J. Anal. Chem.* **2010**, *90*, 671.

20. Askim, J. R.; Mahmoudi, M.; Suslick, K. S. *Chem. Soc. Rev.* **2013**, *42*, 8649.

21. Rakow, N. A.; Suslick, K. S. *Nature* **2000**, *406*, 710.

22. Feng, L.; Masto, C. J.; Kemling, J. W.; Lim, S. H.; Suslick, K. S. *Chem. Commun.* **2010**, *46*, 2037.

23. Suslick, K. S *Curr. Opin. Chem. Biol.* **2012**, *16*, 557.

24. Lin, H.; Suslick, K. S. *J. Am. Chem. Soc.* **2010**, *132*, 15519.

25. Lim, S. H.; Feng, L.; Kemling, J. W.; Musto, C. J.; Suslick, K. S. *Nat. Chem.* **2009**, *82*, 562.

26. Janzen, M. C.; Ponder, J. F.; Bailey, D. P.; Ingison, C. K.; Suslick, K. S. *Anal. Chem.* **2006**, *78*, 3591.

27. Gale, P. A.; Nathalie, B.; Challey, J. E. H.; Louise, E. K.; Isabelle, L. K. *Chem. Soc. Rev.* **2014**, *43*, 205.

28. Martinez-Manez, R.; Ancenon, F. *Chem. Soc. Rev.* **2003**, *103*, 4419.

29. Quyang, G. *Trend. Anal. Chem.* **2012**, *39*, 1.

30. Fabrizzi, L.; Poggi, A. *Chem. Soc. Rev.* **1995**, 197.

31. De Silva, A. P.; Gunaratne, H. Q. N.; Gunnlaugsson, T.; Huxley, A. J. M.; Mc Coy, C. P.; Rademacher, J. T.; Rice, T. E. *Chem. Rev.* **1997**, *97*, 1515.

32. Prodi, L.; Bolletta, F.; Montalti, M.; Zaccheroni, N. *Coord. Chem. Rev.* **2000**, *205*, 59.

33. Valeur, B.; Leray, I. *Coord. Chem. Rev.* **2000**, *205*, 3.

34. Liu, Y.; Tang, Y.; Barashkov, N. N.; Irgibaeva, I. S.; Lam, J. W. Y.; Hu, R.; Birimzahnova, D.; Yu, Y.; Tang, B. Z. *J. Am. Chem. Soc.* **2010**, *132*, 13951.

35. Kaur, K.; Saini, R.; Kumar, A.; Luxami, V.; Kaur, N.; Singh, P.; Kumar, S. *Coord. Chem. Rev.* **2012**, *256*, 1992.

36. Xu, Q.; Li, S.; Cho, S.; Kim, M. H.; Bouffard, J.; Yoon, J. *J. Am. Chem. Soc.* **2013**, *135*, 17751.

37. Ravikumar, I.; Ghosh, P. *Chem. Commun.* **2010**, *46*, 1082.

38. Brooks, S. J.; Gale, P. A.; Light, M. E. *Chem. Commun.* **2006**, 4344.

39. Chutia, R.; Das, R. *Dalton. Trans.* **2014**, *43*, 15628.

40. Gale, P. A.; Busschaert, N.; Haynes, C. J. E.; Karagiannidis, L. E.; Kirby, I. L. *Chem. Commun.* **2010**, *46*, 1082.

41. Liu, B.; Tian, H. *J. Mater. Chem.* **2005**, *15*, 2681.

42. Ajayakumar, M. R.; Asthana, D.; Mukhopadhyay, P. *Org. Lett.* **2012**, *14*, 4822.

43. Wang, Q. -Q.; Day, V. W.; Broman-James, K. *Org. Lett.* **2014**, *16*, 3982.

44. Wang, Q. -Q.; Day, V. W.; Broman-James, K. *Angew. Chem. Int. Ed.* **2001**, *40*, 3066.

45. Chae, M. Y.; Czarnik, A. W. *J. Am. Chem. Soc.* **1992**, *114*, 9704.

46. Tian, T.; Chen, X.; Li, H.; Wang, Y.; Guo, L.; Jiang, L. *Analyst* **2013**, *138*, 991.

47. Lee, K. S.; Kim, H. J.; Kim, G. H.; Shin, I.; Hong, J. I. *Org. Lett.* **2008**, *10*, 49.

48. Song, K. C.; Kim, J. S.; Park, S. M.; Chung, K. C.; Ahn, S.; Chang, S. K. *Org. Lett.* **2006**, *8*, 3413.

YOUR KNOWLEDGE HAS VALUE

- We will publish your bachelor's and master's thesis, essays and papers

- Your own eBook and book - sold worldwide in all relevant shops

- Earn money with each sale

Upload your text at www.GRIN.com
and publish for free